SpringerBriefs in Fire

Series Editor

James A. Milke

For further volumes:
http://www.springer.com/series/10476

Daniel T. Gottuk

Video Image Detection Systems Installation Performance Criteria

 Springer

Daniel T. Gottuk
Hughes Associates, Inc.
Baltimore
21227-1652 MD
USA

ISSN 2193-6595   ISSN 2193-6609 (electronic)
ISBN 978-1-4614-4201-1   ISBN 978-1-4614-4202-8 (eBook)
DOI 10.1007/978-1-4614-4202-8
Springer New York Heidelberg Dordrecht London

Library of Congress Control Number: 2012938704

© Fire Protection Research Foundation 2008
Reprinted in 2012 by Springer Science+Business Media New York

Printed on acid-free paper

Springer is part of Springer Science+Business Media (www.springer.com)

Contents

Chapter 1
Introduction

Computer processing and image analysis technologies have improved substantially over the course of the past decade. This rapidly advancing technology along with the emphasis on video surveillance since 911 has propelled the development of effective video image detection (VID) systems for fire. Fire protection system designers initially employed these VID systems for use in large facilities, outdoor locations and tunnels. However, video-based detection is being used for a broadening range of applications [e.g., 1]. For example, these systems are currently installed in electrical power plants, paper mills, document storage facilities, historic municipal buildings, nuclear research facilities, automotive plants, warehouse/distribution centers, and onshore and offshore oil platforms.

The 2007 edition of NFPA 72, National Fire Alarm Code [2], recognized the use of VID systems for flame and smoke detection. Although recognized, there is limited prescriptive installation and use requirements and there is a general desire by many for the development of performance criteria that ultimately could be utilized for the design of systems or the creation of standards. Since the underlying VID technology and development of standard and network-based camera systems are in a period of fairly rapid advancement [3–5], it is not possible to define a comprehensive set of stand-alone prescriptive requirements. The performance of VID systems depends on both the video hardware and the software algorithms; there is no basic underlying principle, such as there is for ionization or photoelectric detection for smoke detectors. Consequently, performance-based installation and operation requirements are needed.

D. T. Gottuk, *Video Image Detection Systems Installation Performance Criteria*, SpringerBriefs in Fire, DOI: 10.1007/978-1-4614-4202-8_1,
© Fire Protection Research Foundation 2008

1.1 Objective

The objective of this project was to develop fire performance objectives and related criteria for VID systems in selected key applications relevant to their reference in NFPA 72, National Fire Alarm Code. In particular, the goals of the work were to

1. Identify conceptual characteristic fire, smoke and nuisance scenarios for selected applications; and
2. Characterize related performance and installation issues.

The goal of this project was not to develop test methods or listing/approval criteria, but rather to identify the performance issues of concern to the fire safety community.

1.2 Approach

The project goals were achieved through an industry workshop, site visits, data survey and analysis, and experience from the operation and testing of VID systems. Based on the input of the project technical panel, the scope of this project was limited to interior fires at three specified applications: (1) warehouses/distribution centers, (2) large industrial applications, such as petrochemical processing plants and power plants, and (3) atriums. The report first presents a general background of fire VID technology and applicable codes and standards. An assessment of environmental and hazard parameters for the given applications is presented. Based on these parameters and input from the industry workshop, performance parameters and installation issues are discussed.

Chapter 2
Background

2.1 VID Technology

In general, a fire VID system consists of video-based analytical algorithms that integrate cameras into advanced flame and smoke detection systems. The video image from an analog or digital camera is processed by proprietary software to determine if smoke or flame from a fire is identified in the video. The detection algorithms use different techniques to identify the flame and smoke characteristics and can be based on spectral, spatial or temporal properties; these include assessing changes in brightness, contrast, edge content, motion, dynamic frequencies, and pattern and color matching. As an active area of research, there are multiple VID systems in development. However, there are only about five systems that are commercially available. The capabilities of these systems vary from being able to detect only flame or smoke to being able to detect both as well as providing motion detection and other surveillance/security features.

Smoke VID systems require a minimum amount of light for effective detection performance and most will not work in the dark. However, capabilities vary between systems. In general, low light cameras can enhance performance and some systems have been developed to operate in the dark using IR illuminators and IR sensitive cameras [e.g. 3, 4]. Flame VID systems can operate effectively in dark or lit spaces and some systems will have enhanced sensitivity to flaming fires in the dark.

There are two basic architectures utilized by VID systems. Due to limitations of video processing technologies, the initial systems consisted of multiple cameras (usually a maximum of eight) each with an analog cable connection back to a central processing unit that executes all video capture and alarm algorithms (Fig. 2.1). The processing unit typically has relay contact outputs and the ability to send various alarm signals to standard fire alarm control units. Depending on the manufacturer, systems can also record still shots or video clips associated with alarm events and can provide instantaneous video display to a monitor.

D. T. Gottuk, *Video Image Detection Systems Installation Performance Criteria,*
SpringerBriefs in Fire, DOI: 10.1007/978-1-4614-4202-8_2,
© Fire Protection Research Foundation 2008

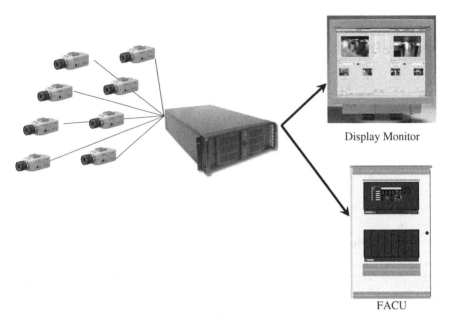

Display Monitor

FACU

Fig. 2.1 VID system with CCTV cameras individually processed by a central control unit that runs the alarm algorithms

Advancement in technologies has allowed the second type of architecture where both the video processing and alarm algorithm execution are performed at the camera in a single, spot-type device (Fig. 2.2), just like a typical optical flame detector. These fire detectors can have onboard storage of video and can be integrated on a closed-circuit system with an additional central processing unit, or it can be integrated as a spot detector on a standard fire alarm system. These devices can also be monitored remotely via network or internet connections. Video events of alarm conditions can be archived for each device and can be displayed automatically to monitors for instantaneous viewing.

As noted, commercially available systems range in capabilities. They also vary considerably in their setup and manner of use. Some systems have little or no user definable settings and are almost plug-and-play, and some systems require a trained manufacturer's representative to customize the system to the application. Once setup, there is little maintenance required of a VID system. Similar to any field of view detector, the primary issue is keeping the optical windows clean and the camera position fixed and unobstructed. Most systems monitor the video image, such as low or exceedingly high light levels, video loss, significant image changes, or obscured camera images, and provide a warning if the image quality is degrading or not sufficient for proper detection performance. Currently, VID systems can be tested/ checked in one of three ways: (1) using a target smoke or fire source, (2) feeding the system a pre-recorded image of a flame or smoke event, or (3) using a product-specific electronic device that directs a pre-set "light" signal to the VID detector.

Fig. 2.2 Example of a spot-type flame VID device with camera and alarm processing in unit (courtesy of Micropack)

VID systems provide unique advantages in a wide range of applications. One advantage these systems offer is the ability to protect a larger area, while still achieving fast detection. This is particularly true for smoke VID systems compared to spot or beam smoke detectors. In many large facilities with excessive ceiling heights, designers find it impractical to use conventional smoke detection devices. VID systems are able to detect smoke or flame anywhere within the field of view of the camera; where as conventional smoke detectors require smoke to migrate to the detector. VID systems can also be used for outdoor applications, such as train stations and off-shore oil platforms.

The ability to use the basic hardware of the VID system (i.e., the cameras and wiring) for multiple purposes is one of the advantages of this technology. Integrating video-based fire detection with video surveillance inherently minimizes certain installation, maintenance and service costs and can increase system availability due to more frequent use of and attention to the video equipment. Providing fire protection for historic buildings poses many challenges to not disturb the historic features of the structure. Running wire and mounting devices of typical fire alarm systems is just not possible in many of these applications for both aesthetic and practical installation reasons. Many museums and historic buildings already have surveillance cameras installed, which makes the use of VID systems attractive.

Another advantage of VID systems is the ability to have live video immediately available upon detecting a pre-alarm or an alarm condition. Immediate situational awareness allows monitoring personnel to easily view the protected area to determine the extent of the fire and to more accurately identify the location. Archiving of still and video images associated with alarm conditions also provides a means of assessing the cause of incidents and also provides a basis for changes in the detection system if the event was a false/nuisance alarm.

2.2 Codes and Standards

The National Fire Alarm Code, NFPA 72-2007, recognizes the use of flame and smoke VID systems. Per the Code, the installation of these systems requires a performance-based design. There are no prescriptive siting requirements. Flame VID systems are classified as radiant energy sensing fire detectors and are treated similar to optical flame detectors. Due to the variability of VID system capabilities and the differences in alarm algorithm technologies,

NFPA 72 requires that the systems be inspected, tested, and maintained in accordance with the manufacturer's published instructions. Appendix A contains excerpts from NFPA 72-2007 that apply to fire VID systems.

Currently, there are no systems that are UL listed, and there is no UL standard that specifically addresses VID systems. Four systems have been FM approved. These include a system that detects only smoke, one that detects only flame and two that detect both. The systems have been approved to meet the requirements of FM Standard 3260, Radiant Energy Sensing Fire Detectors for Automatic Fire Alarm Signaling [6], and UL 268, Smoke Detectors for Fire Alarm Signaling Systems [7].

FM 3260 establishes guidelines for testing detectors per manufacturer-specified sensitivities (i.e., fuel, size, distance and response time). It also requires that the detector sensitivity be established for one or more of the fires below[1]:

- 0.3×0.3 m (12×12 in.) n-Heptane pan fire ~ 126 kW
- 0.3×0.3 m (12×12 in.) alcohol pan fire ~ 27 kW
- 0.3×0.3 m (12×12 in.) JP4 jet fuel fire ~ 146 kW
- 127 mm (5 in.) propane flame from a 0.53 mm (0.021 in) orifice ~ 0.32 kW

Typically, the detector is shielded from the ignition and shuttered once the fire is stabilized. This technique provides more consistency/repeatability in the sensitivity fires. The fires are conducted in large open spaces. If conducted outside, tests are only performed under good weather conditions (i.e., clear, blue sky days). There are no prescribed conditions for indoor tests, but light levels are measured

---

[1] HRR values calculated from empirical data presented by Babrauskas [8].

and documented as part of the approval. Overall, the FM approval is a verification of performance for the tested conditions only.

UL 268 specifies various flaming and smoldering type fires to be conducted in a room that is 10.9 by 6.7 by 3.1 m (36 by 22 by 10 foot high). All of the fires are small. They include pyrolizing wood, a 15.2 by 15.2 cm (6 by 6 in.) wood crib fire, a shredded paper fire, and a 15.8 cm (6.25 in.) diameter pan toluene/heptane fire.

Chapter 3
Environmental and Hazard Parameters

This section describes the foundational information collected for identifying typical/standard fire and smoke scenarios, likely false/nuisance scenarios, and a range of ambient conditions to which VID systems may be subjected in the three target applications. This work included researching and reviewing fire incident data and conducting an industry workshop on VID technology. On-site surveys and interviews with end users were also conducted.

3.1 Statistical Data

The development of performance standards for VID systems relies heavily on the ability to characterize the range of expected fire events in the intended protected spaces. Characterization of these fire events allows for the creation of performance goals that are both realistic and appropriate. A large set of statistical fire data has been compiled through the National Fire Incident Reporting System (NFIRS) and by the National Fire Protection Association via their annual survey of U.S. fire departments. These databases include information about the locations, ignition sources, types of materials first involved, time of day, casualties and damage estimates from actual fire events.

Based on the databases, the following are the available statistical categories that are most closely related to the three target applications of this study:

1. Storage Properties Excluding Dwelling Garages.
2. Industrial and Manufacturing Properties.
3. Petroleum Refineries and Natural Gas Plants.
4. Hotels and Motels.
5. Public Assembly Areas.

D. T. Gottuk, *Video Image Detection Systems Installation Performance Criteria*,   9
SpringerBriefs in Fire, DOI: 10.1007/978-1-4614-4202-8_3,
© Fire Protection Research Foundation 2008

   The most difficult application to specifically extract data for was atriums. The
fire data for hotels and motels and public assembly areas was reviewed since these
occupancies are most aligned to having atriums. Each target application is
addressed below with a review of the applicable incident data.

3.1.1 Warehouses/Distribution Centers

Storage properties included in the statistics include all facilities used for storage of
general items. This includes warehouses and distribution centers used to store
agricultural products such as grain and livestock, refrigerated products, liquid
tanks, or vehicles. These spaces may be used to store flammable materials, be
cluttered and crowded, have large numbers of workers and motion, or have
automated and minimal activities. Figure 3.1a–d shows examples of various
warehouse facilities. It should be noted that residential car garages are not included
in the reported statistics.

   Data is reported for a four year period from 1999 to 2002, in which approximately 22,000 storage facility fires occurred. The number of fires were relatively
evenly distributed over the week, including weekends. The peak time of fires was
between 2 and 5 pm. Table 3.1 displays the material first ignited in storage facility
fires. Storage facility fires are most commonly started either on a structural
member or wall surface (30 %). Another 5 % of fires involve the exterior roof
covering or finish. While the various materials stored in these facilities may create
significant risks for fire escalation and duration, the initiation of a fire event is
more likely to occur on the building structure itself.

   Table 3.2 displays the cause of ignition in storage facility fires. Intentionally set
fires were the leading cause, consisting of 17 % of the fires. Exposure to other fire
(which can be from external exposure or a second item igniting) and electrical
distribution and lighting equipment were the next leading causes (13 and 9 % of
the fires, respectively). Based on property damage, electrical distribution and
lighting equipment was the second leading cause (13 %) after intentional fires
(24 %). The majority of fires (71 %) did not involve equipment in the ignition.

3.1.2 Large Industrial Applications

Even though for large industrial applications it was decided to focus on power
plants and petrochemical processing plants, these categories also represent a broad
range of facilities. At any one particular facility there can be multiple buildings
with different characteristics and functions. Incident data was obtained for
(1) electric generating plants and (2) petroleum refineries and natural gas plants.
For both types of occupancies, the data shows that there are relatively few casualties compared to property damage. For example, an older study of fires in steam,

Fig. 3.1 **a** Example of warehouse/distribution center applications. **b** Example of warehouse/distribution center applications. **c** Example of warehouse/distribution center applications. **d** Example of an automated warehouse/distribution center

heat energy and electric generating plants between 1994 and 1998 shows there were an estimated 362 fires with 1 civilian death, 4 civilian injuries, and $345 million in direct property damage [10]. There were 29 power plant fires from 1990 to 2004 with over $5,000,000 of estimated damage per fire [11].

In electric generating plants for the four year period of 2002–2005, there were 132 structure fires with no deaths and only one civilian injury [12]. The estimated annual average in direct property damage was $8.1 million. The leading equipment involved in ignition was generators (19 % of fires and 54 % of damage) and confined fuel burner or boiler fires, which were 12 % of the fires but only 1 % of the damage. Based on damage, the other two leading equipment were unclassified electrical distribution or power transfer equipment (8 % of fires and 13 % of damage) and conveyors (1 % of fires and 26 % of damage).

Table 3.3 displays the leading materials ignited in structural fires in electric generating plants [12]. Leading materials per the number of fires included

Table 3.1 Top six materials first ignited in storage facilities, annual average events 1999–2002 [9]

Material first ignited	Fires	Civilian deaths	Civilian injuries	Direct property damage (in Millions)
Structural member or framing	3,600 (16 %)	3 (19 %)	14 (7 %)	$66 (13 %)
Exterior wall covering or finish	3,100 (14 %)	0 (0 %)	10 (5 %)	$33 (6 %)
Agricultural crop, including fruits and vegetables	1,900 (8 %)	0 (0 %)	14 (7 %)	$49 (10 %)
Multiple items first ignited	1,200 (5 %)	0 (0 %)	11 (5 %)	$61 (12 %)
Unclassified structural component or finish	1,100 (5 %)	0 (0 %)	6 (3 %)	$21 (4 %)
Rubbish, trash, or waste	1,100 (5 %)	1 (5 %)	8 (4 %)	$19 (4 %)
Total	22,600 (100 %)	17 (100 %)	209 (100 %)	$506 (100 %)

Table 3.2 Leading causes of storage property fires [9]

Cause	Fires	Civilian deaths	Civilian injuries	Direct property damage (in Millions)
Intentional	3,900 (17 %)	3 (16 %)	30 (15 %)	$120 (24 %)
Exposure to other fire	2,900 (13 %)	1 (4 %)	3 (2 %)	$30 (6 %)
Electrical distribution and lighting equipment	2,100 (9 %)	1 (4 %)	22 (10 %)	$68 (13 %)
Smoking materials	1,200 (5 %)	2 (11 %)	12 (6 %)	$18 (4 %)
Lightning	1,100 (5 %)	0 (0 %)	4 (2 %)	$30 (6 %)
Playing with heat source	1,100 (5 %)	3 (19 %)	12 (6 %)	$30 (0 %)
Mobile property (vehicle)	1,000 (5 %)	3 (19 %)	47 (22 %)	$68 (14 %)
Torch (including burner or soldering iron)	700 (3 %)	1 (9 %)	14 (7 %)	$19 (4 %)
Spontaneous combustion or chemical reaction	700 (3 %)	0 (0 %)	5 (2 %)	$16 (3 %)
Contained trash or rubbish fire	400 (2 %)	0 (0 %)	1 (1 %)	$0 (0 %)

flammable or combustible liquid, electrical wire or cable insulation, dust, fiber or lint, and confined burner or boiler fires. Based on damage, the three leading

Table 3.3 Top ten material first ignited in electric generating plant fires, annual average events 2002–2005 [12]

Material first ignited	Fires	Direct property damage (in Thousands)
Flammable or combustible liquid or gas, piping, conduit, hose or filter	16 (12 %)	$482 (6 %)
Confined fuel burner or boiler fire	16 (12 %)	$72 (1 %)
Electrical wire or cable insulation	13 (10 %)	$667 (8 %)
Dust, fiber, lint, including sawdust or excelsior	13 (10 %)	$2,812 (35 %)
Unclassified item first ignited	12 (9 %)	$2,452 (30 %)
Unclassified organic materials	9 (7 %)	$145 (2 %)
Insulation within structural area	8 (6 %)	$57 (1 %)
Exterior roof covering or finish	6 (5 %)	$115 (1 %)
Conveyor belt, drive belt or V-belt	5 (4 %)	$119 (1 %)
Transformer or transformer fluids	5 (4 %)	$1,030 (13 %)
Total	132 (100 %)	$8,153 (100 %)

materials first ignited included the dust, fiber and lint (35 % of damage), unclassified materials (30 %) and transformer or transformer fluids (13 %).

The cause of ignition was classified as confined for 19 % of the fires. Confined structure fires include events such as within chimneys, incinerators, fuel burners or boiler fires. The other two leading categories for cause of ignition were unintentional (39 %) and failure of equipment or heat source (37 %). The last category accounted for 85 % of the damage. There was only one intentional fire per year.

The leading factors contributing to ignition are shown in Table 3.4 [12]. Electrical failure or malfunction produced the greatest number of annual fires and damage (18 and 40 %, respectively). The second was unclassified mechanical failure or malfunction (14 % of fires and 38 % of damage) followed by confined fuel burner or boiler fire (12 and 1 %) and leak or break (9 and 1 %). The primary heat for the ignition sources was from radiated or conducted heat from operating equipment (20 % of fires and 34 % of damage) and unclassified heat from powered equipment (18 and 28 %).

Based on this incident data, the majority of electric generating plant fires causing the most damage were a result of failure of equipment or a heat source and involved generators, electrical distribution or power transfer equipment, and conveyors. The initiating fires primarily involved dust, fiber and lint, unclassified materials, and transformer or transformer fluids. The occurrences of fires at electric power plants were fairly uniformly distributed throughout the week with a notable decrease on Sundays. During the day there was a peak in the number of fires during normal working hours between 9 am and 6 pm.

Statistics for petroleum processing facilities and natural gas plants was obtained from Ref. [13]. Between 1994 and 1998, there were only about 228 fires or

Table 3.4 Leading factors contributing to ignition in electric generating plant fires, annual average events 2002–2005 [12]

Factor contributing to ignition	Fires	Direct property damage (in Thousands)
Electrical failure or malfunction	24 (18 %)	$3,293 (40 %)
Unclassified mechanical failure or malfunction	18 (14 %)	$3,067 (38 %)
Confined fuel burner or boiler fire	16 (12 %)	$72 (1 %)
Leak or break	12 (9 %)	$56 (1 %)
Heat source too close to combustibles	7 (6 %)	$40 (0 %)
Failure to clean	7 (5 %)	$1 (0 %)
Cutting or welding too close to combustibles	7 (5 %)	60$ (1)
Total	132 (100 %)	$8,152 (100 %)

Table 3.5 Top six types of material first ignited in petroleum and natural gas, plant structure fires, annual average events 1994–1998 [13]

Type of material	Fires	Civilian deaths	Civilian injuries	Direct property damage
Flammable or combustible liquid	18 (40.9 %)	0 (NA)	2 (87.4 %)	$2,261,000 (87.1 %)
Volatile solid or chemical	7(17.1 %)	0 (NA)	0 (0 %)	$71,000 (2.7 %)
Wood or paper	5 (11.7 %)	0 (NA)	0 (0.0 %)	$245,000 (9.4 %)
Gas	5 (10.9 %)	0 (NA)	1 (25.1 %)	$5,000 (0.2 %)
Material compounded with oil	2(5.7 %)	0 (NA)	0 (0.0 %)	$2,000 (0.1 %)
Natural product (e.g., rubber)	2 (4.6 %)	0 (NA)	0 (0.0 %)	$5,000 (0.2 %)
Total	44 (100.0 %)	0 (NA)	3 (100.0 %)	$2,597,000 (100.0 %)

explosions at these facilities per year. Sixty four percent of the fires occurred outside, with 40 % of these fires involving property of value. Structure fires accounted for only one-fifth of the fires, but resulted in more than half of the dollar loss for all fires at these facilities. Annually there were only 44 structure fires on average. These fires resulted in an average of three civilian injuries per year and an annual average of $2.6 million in direct property damage. This data shows the need for detection and rapid response to these fires.

In petroleum refineries or natural gas plants, 25 % of the structure fires per year began in process or manufacturing areas. This was the leading area of origin, accounting for 79 % of the direct property damage. The second leading area of origin was product storage area, tank or bin, which represents 12 % of the fires and 10 % of the damage.

Table 3.6 Top six ignition factors in petroleum and natural gas plant structure fires, annual average events 1994–1998 [13]

Ignition factor	Fires	Civilian deaths	Civilian injuries	Direct property damage
Part failure, leak or break	14 (31.5 %)	0 (NA)	0 (0.0 %)	$1,544,000 (59.5 %)
Unclassified or unknown-type mechanical failure or malfunction	5 (11.1 %)	0 (NA)	0 (15.3 %)	$65,000 (2.5 %)
Cutting or welding too close	4 (10.2 %)	0 (NA)	0 (0.0 %)	$4,000 (0.2 %)
Unclassified or unknown-type operational deficiency	3 (6.9 %)	0 (NA)	1 (25.1 %)	$49,000 (1.9 %)
Combustible too close to heat	3 (6.2 %)	0 (NA)	0 (0.0 %)	$9,000 (0.4 %)
Lightning	2 (5.3 %)	0 (NA)	0 (0.0 %)	$7,000 (0.3 %)
Total	44 (100.0 %)	0 (NA)	3 (100.0 %)	$2,597,000 (100.0 %)

Accelerant, gas or liquid in or from pipes or containers and fuel was the first form of material ignited. These events accounted for 54 % of the fires and 89 % of the damage. Table 3.5 presents the type of materials first ignited in the structure fires. As expected, flammable or combustible liquids were the leading material, constituting 41 % of the fires. Volatile solid or chemicals accounted for the next largest group (17 % of the fires), followed by wood or paper (12 %) and then gas (11 %). Table 3.6 shows the top six ignition factors for the structural fires. The leading factor was part failures, leaks or breaks (32 % of fires and 60 % of the damage). Other unclassified mechanical failures or malfunctions was the second leading factor (11 %) followed by cutting or welding (10 %). Only 1 % of the fires were incendiary or suspicious.

3.1.3 Atrium

Atriums are large open spaces with high ceilings commonly found in shopping malls, hotels, office buildings, or airports. These spaces may have long viewing distances and have large numbers of people moving about within the protected space. The NFPA has not compiled statistical data for atria fires specifically; however, several documented fire events have occurred. All major events that have occurred in an atrium involve smoke entering from a fire developing in an adjacent space. There were no major fire events identified that initiated within the atrium. Many atriums are equipped with smoke control ventilation systems, the primary reported events are when ventilation systems fail and the atrium filled with smoke.

3.2 Industry Workshop

On March 10, 2008 an industry workshop on VID Systems was held in Orlando, Florida. There were 50 attendees representing manufacturers, fire protection engineers, end users, authorities having jurisdiction, standards organizations, and insurance carriers. A complete list of attendees is included in Appendix B. The purpose of the workshop was to solicit input and seek consensus of key environmental issues, sources and performance parameters for the use of VID systems in the selected occupancies. After presentations on the general background of VID technology and a summary of the statistical data presented in Sect. 3.1, the attendees were divided into three groups corresponding to the selected occupancies of warehouses/distribution centers, large industrial applications, and atriums. Each group was given a set of questions and requested to discuss and come to consensus on the answers. The questions are included in Appendix B along with the consolidated group responses. The set of questions was the same for each occupancy and focused on the purpose and objectives of the VID systems for these occupancies, the desired system requirements, and site attributes.

The general consensus responses for each application are presented below. One clearly apparent conclusion is that the design of a VID system requires a performance-based approach as is also noted in NFPA 72. Although, there was a general consensus on a number of application questions, the discussions plainly pointed out that depending on the particulars of the application, the installation goals and the installation criteria could vary significantly.

3.2.1 Warehouses/Distribution Centers

For warehouse/distribution centers, the main objective for using VID systems was stated as property protection and mission continuity, particularly via automatic suppression activation. Besides detection for activation, the use of VID systems was deemed useful for verification of zone evacuation purposes, especially as part of a mass notification system. The use of smoke detection was not a code requirement for warehouse and distribution centers, but was considered to be necessary for high-end storage (e.g., pharmaceuticals and expensive electronics where smoke damage is critical). The primary use of VID for this application was flame detection; however, as will be noted below, there were limitations to this conclusion.

Although the consensus was that the primary application of VID technology in warehouse and distribution centers is flame detection for automatic suppression, the wide range of warehouse designs and commodities made it difficult for the group to define design fires and installation criteria with certainty. Widely ranging characteristics of these facilities include aisle widths, storage heights, and types of storage (e.g., cold and automated). For instance in automated rack storage facilities with automated machinery traversing down aisles (Fig. 3.1d), the environment can be very dangerous for people and present very limited fields of view for VID systems. Therefore, many automated warehouse sites would not send people into fight a fire; so defining fires for early warning smoke detection was not seen as useful or practical. Additionally, the utility of flame detection is uncertain due to potentially limited fields of view in narrow aisles. Despite such examples and the group's understanding of needing research to more accurately define parameters, the group stated that VID systems should detect 20–100 kW fires with smoke algorithms and larger fires on the order of 300 kW for flame alarms. However, the there was uncertainty in stating the 300 kW fire.

For many warehouse and distribution centers, the spaces are large and fairly open with up to 40 ft ceilings and greater than one million square feet. VID protection above aisles was noted as preferable. However, this would be more appropriate for smoke detection not flame, particularly for spaces with high storage heights. It was recognized that camera locations is highly dependent on the size and type of warehouse and the type of fire to be detected. Facility attributes, such as vehicle and personnel movement and changing background, color and objects, can impact the placement and performance of VID systems. One example is the use of high volume, low speed (HVLS) fans that can be 24 ft in. diameter. Besides the physical obstruction these HVLS fans can impose, they also move large amounts of air that would move and substantially dilute smoke from fires; this can make detection more difficult. Consequently, a fire size that is detectable in one facility without fans may not be detectable in a facility with HVLS fans.

The group preferred to have VID devices that were self-contained where both the video processing and alarm algorithm execution are performed at the camera. Integration with other building, security and fire alarm systems is desirable as long as all components are reliable and listed/approved. No specific listing/approval requirements were established other than a general claim that the process establishes reliable performance. It was acknowledged that current listing and approvals for VID systems could be in the form of a special application as opposed to meeting just one specific standard for VID systems.

A number of false/nuisance sources were noted for the warehouse/distributions centers. Some, such as movement of people and machinery and fans, have been noted above. Others included change of ambient light, dust, steam, obstructions, condensation, space heaters, sparks from controlled fire, and exhaust from forklifts. Changes in ambient light could include the opening and closing of large outside doors. Since many of the facilities operate throughout the night, changes in artificial lighting was not a concern in those applications.

The group concluded with a statement that the best application for VID in warehouse/distribution centers were those for high-end and electronic storage. These applications would benefit from both smoke detection for early warning as well as from flame detection.

3.2.2 Large Industrial Applications

Large industrial applications cover a wide range of facilities and processes. Even within the narrower categories of power plants and petrochemical processing plants that were focused on, there is still a broad range of detection environments. In general the workshop group agreed that mission continuity and property protection (in that order) were the primary reasons for using VID systems in large industrial sites. (The statistical data presented in 5.1.2 supports the working group consensus relative to the life safety versus property damage impact of fires in these facilities.) Although life safety is an important system objective, it was deemed more of a future need. Many industrial facilities have areas with limited or frequently no personnel. Combining VID fire detection with surveillance was considered a plus. In these applications, a recurring point was that detection systems are desired to provide early warning necessary for quick, manual intervention; toward this end, both smoke and flame detection is needed. VID systems are desired for improved operator response and positive event verification.

Similar to the warehouse applications, the large diversity in industrial applications made the selection of specific fire hazard and detection criteria difficult. Consequently, it was recognized by the group that varying site specific conditions would dictate a performance-based approach to detection system design. Class A and B fires were considered typical and Class C and D fires were considered as possible scenarios. Outside applications were considered just as important as interior fire scenarios. Examples include coal pile storage and switch yards.

The group did not specify quantitative fires sizes to be detected, but noted that for the desired goal of quick manual intervention, there is a need for detecting the very early incipient stage. In addition, the group indicated the desired detectable fire was 7×7 pixels of the video image. This generally represents the smallest size fire detectable by current VID systems. However, the actual fire size or area of smoke represented by 7×7 pixels of video image is dependent on the field of view set by the camera lens. As noted earlier, the group realized that establishing fire sizes and sensitivity settings is application specific and cannot be quantified without a performance-based analysis.

Integrating VID systems with fire alarm systems and other building/security systems is desired. Although it was believed that many applications would require a listing for the VID system, this was not always the case for many industrial applications. In general the following equipment/system requirements were noted as having priority:

1. Self-supervision.
2. Secondary power.
3. Camera/signal integrity.
4. Open signal architecture to work with other systems.
5. Ability for acceptance testing and on-going testing (recognized as important but not specified as to how it should be done).

The capability of cameras to pan, tilt and zoom was not deemed a necessity, but was viewed as a future advanced feature. Cameras should be mounted as would typical surveillance cameras to provide generally horizontal views and accessibility for maintenance and testing.

The predominant concerns for false/nuisance sources that were noted included dirt, oils, humidity and light. It was recognized that false/nuisance sources could be managed by video verification. The operation of equipment and vehicles and the movement of people within the field of view of a camera are common events. Due to the varied sizes and layouts of industrial spaces, the VID systems need to be flexible in their fields of view and be able to work in large open spaces to smaller or congested spaces. Ventilation was deemed an important issue since many spaces have large air flow rates. A primary concern of high air flows is dilution of smoke, which will make detection more challenging.

3.2.3 Atriums

The working group assumed that most atriums include the presence of people. Consequently, life safety was the priority objective for a VID system (i.e., to facilitate egress). Security was stated as the second objective with mission continuity last. The group believed that a VID system was only helpful for fires in the actual atrium, as opposed to connected areas which typically have spot smoke detection systems. There was much discussion on how VID systems would integrate into existing building fire protection systems for buildings with atriums. The consensus was that VID systems would primarily provide a positive alarm sequence and secondly be used to activate smoke control and initiate manual response (i.e., the fire brigade). The positive alarm sequence consists of receiving pre-alarm signals that would initiate an investigation period in which the alarm is verified. This requires the VID system to provide video images for manual review, which allows for discernment of false/nuisance alarms and judgment calls for response based on the specific scenario. Many atriums are required to have smoke control systems. These systems typically are activated on overhead sprinkler activation or smoke detection (in many cases beam smoke detectors). As part of the overall system, activation notification is provided at a fire alarm control panel. A VID system would need to be able to integrate with the smoke control systems and fire alarm control panels. Using the VID system for security purposes was seen as permissible as long as fire detection and alarm was a priority function and not affected by the security features.

For the atrium applications, VID systems would have to be listed/approved for smoke detection. The need for flame detection was judged to be dependent on the specific needs of a building application, but was not deemed necessary as was smoke detection. The primary fire type expected is Class A. Similar to the other occupancies, the group stated that a prescriptive design fire size could not be given and that the target fire needs to be determined based on an engineering analysis for the subject space which would include an assessment of hazard and risk. It was stated that the fire must be discernable to viewers, indicating that to have positive alarm verification, the smoke or fire must be visible in the video image. This does imply a minimum source size since VID systems can detect smoke in the video image that is not readily noticeable by a person looking at the same image. However, the amount of smoke is dependent on the camera settings, the field of view, the ventilation, and the fire source location relative to the camera. Consequently, it is not possible to practically quantify the source based on this criteria alone.

One installation concern was mentioned that may impact detection. Cameras would need to be located to provide semi-overlapping coverage to prevent blind spots under a particular camera. There was a concern that a large fire under a camera could obscure the camera view without an alarm condition; that is, the sudden obscuration would be interpreted as a nuisance source. The group did not see the need for pan tilt zoom camera capability.

3.3 End User and Site Information

Several occupancies were evaluated to provide additional guidance to the range of operating conditions that VID systems would need to operate within as well as to assess the target fires to be detected. One facility included a petrochemical processing site with a range of buildings and processes. At this one site, the areas to be protected by detection ranged from small congested rooms to very large, predominantly open spaces. The primary user requirement was to detect flame, not smoke. The desired field of view of the detectors ranged from as small as 10–20 feet to upwards of 200 feet. Environmental conditions ranged from sub-zero temperatures to over 100 °F. The spaces were relatively sparsely manned with little visible activity.

Figure 3.2a–d show photographs of some of the environments encountered. In many instances, there were considerable obstructions within the detector field of view, such that a flame would not be fully visible to a camera. For example, relative to the cameras field of view, fires could be behind an array of piping, such that multiple slices of the flame would be seen. In some cases a large portion of the fire could be behind a valve or tank so that the camera would only see the top or bottom half of the flame. Based on these observations, it became obvious that a VID system would need to be able detect fires that are not fully within the camera field of view. This is noteworthy in that most testing conducted for VID and even optical flame detectors consists of flames fully in the open within the detector field of view.

Besides obstructions, it became apparent that fire scenarios could likely develop with flames impinging on piping, tanks or equipment. A breach in a seal or pipe connection could lead to a spray or jet fire that immediately impinges on adjacent objects. Similarly, pool fires developing under valves or pipes would be deflected

Fig. 3.2   a Example of a fairly open field of view in a petroleum processing plant. **b** Example of a petroleum processing plant. **c** Example of a petroleum processing plant. **d** Example of a petroleum processing plant

laterally. These scenarios can result in distorted flame shapes and non-classical fire plumes that are quite different from fires burning in the open, on which some VID systems are developed. Algorithms that compare visual patterns to training sets could be challenged by distorted flames and broken fire plumes. There has been limited work done in this area. Primarily shielded 1–2 MW diesel fires conducted in the Lincoln Tunnel were not able to be detected by several installed VID systems [14]. In the open, there is no doubt that the fires were detectable by the systems.

The potential fire hazards existing at the petroleum processing site include:

- Crude oil pools and sprays.
- Triethylene glycol (TEG) pools and sprays.
- Diesel fuel pools and sprays.
- Methanol pools and sprays.
- Natural gas jets.
- Electrical motors, switch gears, battery casings, and cable tray wiring.
- Transformers.
- Lube oil pools and sprays.
- Cellulose materials.

There are an infinite number of possible fire scenarios that could occur, including a range of pool sizes, different spray patterns, and various flame geometries due to fires burning in and around multiple piping and equipment configurations. The site operator recognized that it is important that a detection system be able to alarm to both the minimum event of concern that could result in plant damage and a rapid, large event that may saturate the detectors. There is a concern that a rapid, large fire may result in no fire alarm because the event is perceived as a nuisance source, such as the opening of a large outside door with sunlight flooding in. A detector capable of producing an acceptable alarm for the smallest identified threat scenario and for a large, instantaneous fire was expected to satisfy the range of anticipated fire scenarios.

However, determining the smallest threat scenario is challenging as it depends on many circumstances. In general, damage to personnel and property from a fire event is due to the heat flux at the specific distance of the target from the flame. It has been stated by Babrauskas for liquid pool fires that the potential for destruction and escalation exists when the pool has an effective diameter greater than 0.2 m (7.9 in.) [8]. At this pool size, the fire enters a mode where the radiative heat transfer becomes significant. A crude oil fire of this size would yield a total heat release rate (HRR) of about 46 kW. Using the Babrauskas estimate assumes the main threat is radiated heat. Smaller fires could be a threat if the flame is impinging (convective heat transfer) on a component that could lead to further fuel release and fire escalation. Identifying the smallest fire threat appropriate for establishing the lower detection limit can also be dependent on risk analyses that take into account probability of occurrence and consequences rather than just assessing what fire can do a minimum threshold level of damage. Consistent with the FM 3260 standard, the site operator deemed a 0.3×0.3 m (12×12 in.) fire as representing the smallest fire size needed for detection performance. This fire size is slightly larger than the 0.2 m (7.9 in.) minimum fire claimed by Babrauskus to be capable of causing escalation and damage. However, it is still a relatively small fire, representing about 100 kW with an approximately three to four foot flame height.

Based on on-site concerns, the following sources were identified as potential false/nuisance sources: heat/light sources, arc welding, torch cutting, sparks from grinding metal, and general personnel operations. There were limited doors and windows in most of the site buildings and structures. Therefore, changing outdoor

Fig. 3.3 Example of a turbine building at a coal-fired electrical power plant

lighting or bright sun exposures inside the facilities were not a significant concern. Besides the detection performance requirements, the flame detection system was required to be integrated into a site-wide control and monitoring system. The ability to have video was not necessarily a priority. The main objective was to have reliable and fast detection coverage over the wide range of fields of view.

A second facility that was evaluated was a coal-fueled electric power plant. The site operator was primarily interested in fire detection in a large turbine building (Fig. 3.3) and coal handling spaces (Fig. 3.4). The main objective of the detection system was operation continuity and property protection. The use of video for security/surveillance was not an issue or requirement. The site desired to have both flame and smoke detection with access to video for visual assessment of the space during detected events. The desire is to have early detection of small flaming fires and early stages of smoldering sources. As such, evaluation of a 23 × 33 cm (9 × 13 in.) pan, isopropyl alcohol fire (∼25 kW) was deemed representative for an acceptable detection limit. In addition, white and black smoke sources were used to evaluate the system. Figure 3.5 shows an example of smoke detection on the turbine deck.

In the coal handling areas it is desired to pick up fires as small as possible. Besides fire detection, the smoke alarm algorithms were also deemed important for detecting coal dust. So contrary to some applications, where dust may be considered a nuisance source, it was a desired detection event at this plant. Expected events to be detected include lube oil, coal, conveyor belts and cable fires. Although preferred, the detection equipment is not required to be listed or approved.

Fig. 3.4 Example of a coal handling room at a coal-fired electrical power plant

Fig. 3.5 VID system detecting smoke in a power plant (courtesy of axonX)

Similar to the petrochemical application, the areas needing detection coverage frequently are lightly manned with minimal activity. Figure 3.3 shows a photograph of the turbine building. This space is primarily large and open except for the turbine units and other sizeable tanks, equipment and a modular office space structure. The main space communicates with adjacent spaces and the floor below (see opening in floor in Fig. 3.3). Although operations do not normally entail a lot of movement on the turbine deck, there are various activities that occur. One operation to be considered for VID system setup is the use of the overhead crane that traverses the length of the building. This can present a significant obstacle as well as a potential source for creating false alarms. Another attribute of the space to be considered is the large amount of windows both high and low in the space. An installed VID system at the site masked out the windows from the video image detection zones. Other than the light through the windows, most of the background color is gray and brown.

The coal handling spaces can be quite different from the turbine building in that there is moving equipment, dust generation, and narrow fields of view. For example, the space shown in Fig. 3.4 includes a rubber conveyor belt to transport coal. The room is very long and narrow with relatively shallow ceilings. The walls are white and there were banks of windows down one side of the space. This particular space shown had a large fan at the end of the room that created large air flow rates through the space to keep it cool. The high air flow also challenges smoke detection by quickly diluting and diffusing the smoke plume from the source. It was noted that most coal handling rooms did not have such ventilation. Also, all white walls in these areas are washed down every day.

The wash down procedures pose a potential false/nuisance source. A number of other false/nuisance sources were identified by the site operators. One included the use of acetylene torches for maintenance and repair operations. Another was steam (as seen in Fig. 4.1). However, similar to dust generation, steam plumes within this facility are associated with abnormal operation and are initially desired to be detected. In general, personnel would accept one false/nuisance source per day for the site, regardless of the number of cameras.

Information was also obtained regarding the use of a VID system in a warehouse distribution center. The owner did not set any specific goals for the performance of the detection system. Instead, the capabilities of the VID system were demonstrated and accepted as beneficial for providing early warning with the general goal of maintaining business continuity and the protection of property. Early warning detection was desired for on-site fire fighter response before sprinkler activation. After installation, the benefits of using the video system for security and surveillance proved more worthwhile than initially expected. There were no code requirements to have a detection system.

In the facility, most cameras are mounted 20–30 ft high and are tilted down approximately 20–25 degrees from horizontal to capture ceiling to floor views. An example of a camera view is shown in Fig. 4.2. Background images in a camera view change. They can be open rack shelving or shelves filled with varying packaged materials. Loading and unloading areas can change relatively quickly with palletized commodities. There is a lot of activity in the distribution center, including people and vehicles. The owner stated that they would accept two false/

nuisance alarms a day for 100 % coverage of the distribution center. The video is monitored 24/7 by security personnel who have oversight of other systems such as a mass notification system.

Chapter 4
Discussion

The industry workshop on VID systems, as well as a review of fire incident data and multiple occupancies that are using or considering the use of VID systems, have clearly revealed the complexity of fire VID applications. Even for similar occupancies or even different camera locations within the same space, there can be large variations in the environmental factors that impact VID performance. In general, background images (e.g., structures and equipment), colors of walls and objects, lighting, and activities within the field of view of a camera can have effects on the operation of a system. This variation does not inherently imply problems for VID systems, but it does create a problem for selecting specific environmental parameters in developing performance criteria that are to be broadly applicable.

The performance of a smoke or flame VID system must take into account three general items:

1. Fire sources.
2. Environment.
3. System variables.

In general, fire sources are defined by the specific application being addressed and the desired response from detection to manual or automatic actions. The environment, such as background, foreground, lighting, contaminants, etc., are dependent on the building and its operations. System variables are dependent primarily on the specific VID system being used; although some aspects, such as field of view, are also dependent on the environment being protected. These general items are discussed separately below (Figs. 4.1 and 4.2).

D. T. Gottuk, *Video Image Detection Systems Installation Performance Criteria*,
SpringerBriefs in Fire, DOI: 10.1007/978-1-4614-4202-8_4,
© Fire Protection Research Foundation 2008

Fig. 4.1 Example of a potential nuisance source—steam rising from space below

Fig. 4.2 Example of warehouse/distribution center applications (courtesy of axonX)

4.1 Fire Sources

The selection of fire sources is dependent on the hazards at a specific site and the goals of the global fire protection systems for that site. Inherently, due to the variability in these dependencies, selecting a specific set of fires that fits all applications is not possible. However, based on the fire incident data along with input from designers and end users, some general classifications can be made for each of the occupancies studied.

4.1.1 Warehouse/Distribution Centers

Storage facility fires are most commonly started either on a structural member or wall surface (30 %) and the majority of fires (71 %) did not involve equipment in the ignition. The leading cause (17 %) was intentional set fires. In addition to quickly detecting such fires, VID systems can provide a useful video log that can potentially be used to identify and prosecute the fire starter. By their nature these facilities contain a wide range of commodities which make it difficult to define generic design fires for any detection system. Despite the workshop industry group's understanding of needing research to more accurately define parameters, the group stated that VID systems should detect 20–100 kW fires with smoke algorithms and larger fires on the order of 300 kW for flame alarms. Some large distribution centers desire detection as early as possible for on-site fire fighter response before sprinkler activation. To this end, fires in the range of 20–100 kW would be appropriate to give adequate response time before fires grow and potentially actuate sprinklers. For example a 100 kW fire will grow to 1,100 kW in 3.5 min if developing as a medium growth rate fire, which is a reasonable, if not slow, growth rate for a warehouse [2]. Class A and Class B fires would be appropriate for flaming fires in these facilities, Class A probably being predominant. For smoldering fires, Class A commodities and electrical wiring and cables would be relevant sources.

4.1.2 Large Industrial (Petrochemical and Electric Power Plants)

Based on the incident data, the majority of electric generating plant fires causing the most damage were a result of failure of equipment or a heat source and involved generators, conveyors, and electrical distribution or power transfer equipment. The initiating fires primarily involved dust, fiber and lint, unclassified materials, and transformer or transformer fluids. In petroleum refineries and natural gas plants, accelerant, gas or liquid in or from pipes or containers and fuel was the first form of material ignited; these events accounted for 54 % of the fires and 89 % of the damage.

Although there is some overlap, even for these two types of industrial facilities, there is significant diversity in the types of fires that are expected to occur. Consequently, it was recognized by the workshop industry group that varying site specific conditions would dictate a site specific, performance-based approach to detection system design. Class A and B fires are considered typical and Class C and D fires were considered as possible scenarios. The group did not specify quantitative fires sizes to be detected, but noted that for the desired goal of quick manual intervention, there is a need for detecting the very early incipient stage. This approach agrees with one petrochemical facility where the site operator deemed a 0.3×0.3 m (12×12 in.) pool fire as representing the smallest fire size needed for detection performance. This size is consistent with the FM 3260 standard, representing about a 100 kW fire with an approximately three to four foot flame height.

The operators of a coal-powered electric plant desire to have early detection of small flaming fires and early stages of smoldering sources. Particularly in the coal handling areas it is desired to pick up fires as small as possible. Based on testing performed, fires as small as about 25 kW were deemed appropriate for detection. This is consistent with the lower limit proposed by the warehouse workshop group.

In general, Class B fuels would be most appropriate for these industrial occupancies, including hydraulic fluid, diesel fuel, and transformer and lube oils. Flammable gas and, to a lesser extent, Class A combustibles are also quite appropriate. Since many of the fires in these two industrial occupancies are a result of mechanical failures/malfunctions and release of gas and liquids from pipes and containers, it is expected that fires can start as jets, sprays, and pools. These fires can result in distorted flame shapes and non-classical fire plumes that are quite different from fires burning in the open.

4.1.3 Atriums

The NFPA has not compiled statistical data for atria fires specifically; however, several documented fire events have occurred. There were no major fire events identified that initiated within atriums. All major events that have occurred in an atrium involve smoke entering from a fire developing in an adjacent space. For these scenarios, sources would generally consist of larger Class A fires. Likewise, fires developing in the atrium will primarily be Class A, ranging from common furniture arrangements and kiosks to seasonal displays. The workshop atrium group stated that a prescriptive design fire size could not be given and that the target fire needs to be determined based on an engineering analysis for the subject space. For the design of smoke management systems in atriums, approximately 2 MW fires are used for activation [15]. If the goal of the system is to activate smoke management, than a 2 MW Class A fire would be an appropriate upper bound. However, the workshop group identified the need of the VID detection system to provide a pre-alarm signal to initiate an investigation period prior to further actions of people evacuation and activation of smoke management. They also specified that the fire must be discernable to the viewer. Depending on multiple factors (e.g., fire type, ventilation and obstructions) a discernable fire could imply tens of kilowatts to hundreds of kilowatts.

Particularly for detection to provide early notification as part of a positive alarm sequence before people evacuation and activation of smoke management, a Class A fire of approximately 100 kW would be a reasonable performance objective given the size and openness of most atriums. It is important to note that this size fire would not be detectable by most sprinkler and smoke detection systems in these spaces and may represent a very conservative small fire size. Ultimately, a design fire must be selected based on an analysis of the hazard and potential consequences relative to the available fire protection systems in the specific atrium as well as the personnel and emergency plans in place. For example, the desired detection fire size may depend on the training of on-site personnel and whether the fire department is automatically notified upon alarm.

4.1.4 Summary of Fire Sources

For warehouse/distribution centers and large industrial applications, such as power plants and petroleum processing plants, the main goal of VID systems is to provide property protection and mission continuity. In general, the fire detection systems are desired to provide early warning necessary for quick, manual intervention. The primary concerns for these occupancies are Class B and Class A fuels; however, electrical failure and electrical distribution fires constitute a significant number of incidents in both of these general applications and are pertinent to early warning detection, particularly for smoke production. Especially for the industrial applications, combustible and flammable liquids and gases are a concern. For both types of occupancies, the goal of early detection leads to the performance objective of detecting fires on the order of 25–100 kW.

The same argument for fire size can be made for atriums, for which life safety is the priority goal. Since atriums routinely include the presence of people, providing early detection for evacuation is a main objective. A Class A fire of approximately 100 kW would be a reasonable performance objective given the size and openness of most atriums. The minimum fire size for the activation of smoke management systems in atriums to maintain tenable conditions within the space and adjacent connected spaces will be larger than 100 kW. Therefore, a 100 kW fire provides time for notification of occupants prior to when a smoke management system would operate, after which, by design, occupants should still have time to evacuate.

4.1.5 False/Nuisance Sources

The detection of fires is integrally related to the ability of the system to discriminate false or nuisance sources. A detection system that detects 100 % of all fires, but also has numerous false/nuisance alarms will not be practical nor accepted. Identifying and standardizing false/nuisance sources is even more difficult for VID systems than for spot smoke detectors and OFDs. This is due to an infinite array of conditions that can affect VID systems and the fact that differences in VID algorithms (i.e., product specific) will cause the different systems to react differently to environmental parameters. For example, two VID systems may be able to ignore a flashing light on a forklift truck when viewed directly, yet one system may alarm when the truck passes behind an open frame rack assembly, whereas the other does not because of its different algorithms. There is not as much variation between the same type of spot detectors (photoelectric for example) from different manufacturers relative to potential false/nuisance sources. For example, most will false alarm when exposed to steam, regardless of other room parameters (e.g., lighting, wall color, structural/content visual patterns). The bottom-line is that VID system alarm algorithms are dependent on more external variables than spot smoke and optical flame detectors.

Since VID systems are able to provide immediate situational awareness to monitoring personnel via video, there appears to be a higher acceptance of false/nuisance alarms by site operators. Operators of both a warehouse/distribution center and the electric power plant stated that one false/nuisance source per day for the site, regardless of the number of cameras would be acceptable. The distribution center was even okay with two false/nuisance sources per day.

Video archiving of events provides a means to diagnose potential recurring false/nuisance problems and a basis to make system adjustments. Some systems have the ability to ignore areas of the field of view that may have potential false/nuisance sources, to adjust sensitivity and to adjust the persistence time of the event before an alarm signal is issued. Specific alarm algorithms have also been developed by manufacturers to avoid common nuisance events.

In developing a performance requirement for false/nuisance source discrimination, a highly variable set of parameters including the source as well as environmental characteristics need to be defined. The variability is greater than that for defining false/nuisances sources for other smoke and fire detection systems, for which there is also little in the way of standardized false/nuisance testing. That being said, a number of potential nuisance sources have been identified for VID systems. In general, motion of any object that may have visual attributes similar to a flame or smoke can present a potential for being a false alarm. The ability to deal with this broad statement is highly dependent on the VID manufacturers alarm algorithms. Other concerns for false/nuisance sources are heat/light sources, arc welding, torch cutting, sparks from grinding metal, and general operations, such as movement of people and machinery and fans. Change of ambient light, dust, steam, obstructions, and lens contamination from dirt, oils, and condensation are other possible sources. Some of these potential sources are addressed within the UL 268 and FM 3260 test standards and are included as part of the listing and approval of other detection systems. For examples, optical flame detector manufacturers typically test and define the range of immunity to welding.

4.2 Environment

The environment, such as background, foreground, lighting, contaminants, etc., is dependent on the building and its operations. The primary issue is how do the environmental parameters affect the image within the field of view of the VID camera and how do these effects impact the alarm algorithms. This section will primarily address how the environmental parameters affect the image. The impact on alarm algorithms is highly dependent on the specifics of each manufacturers system and therefore cannot be addressed fully in detail, but only in a general manner.

Even within the three applications considered in this study, there is a myriad of environmental images that a VID camera would view. These range from large, fairly open spaces to small spaces and highly obstructed views. The spaces can be routinely void of personnel and equipment moving around, or they can have almost continual motion, such as the loading area of a distribution center. The two primary issues that arise from these ranges of conditions are obstructions in the field of view and variable motion. It is almost impossible to define more specific performance requirements for motion within a field of view other than stipulating that a system must be nearly free of false alarms for the application. Demonstrating compliance with this requirement can obviously come via installation in the specific application, but it can also be achieved from historical performance from similar installations. To a very limited extent, standardized tests can be performed to assure a minimal level of false alarm resistance to motion in the field of view. However, it is expected that such standardized testing would be of marginal benefit than is achieved by simple customer demand. If a system is not robust enough to operate in typical environments without numerous false alarms, it will not succeed as a viable commercial product.

Relative to obstructions, there is a lack of data and testing of fire detection capability (particularly for flame detection) for fires that are partially obstructed in the field of view or flames impinging on objects. For example, flames impinging on piping, tanks or equipment can have distorted flame shapes and non-classical fire plumes that may look quite different in a video from fires burning in the open. As can be seen in Fig. 3.2a and b in industrial facilities, fires burning behind multiple items relative to the camera field of view could appear as multiple, discontinuous segments. Therefore, for many applications, VID system performance should stipulate whether fires are to be seen directly or with obstructions within the field of view. Secondly, even if a fire is to be viewed directly, the performance of the system should specify whether the fire may be distorted due to impingement on equipment. This latter issue may only be significant for smaller size fires, because as a fire grows larger, the scale of the total fire size will exceed the scale of the initial flame distortion.

Variable lighting within a space can have adverse effects on VID system performance. All artificial and natural light sources should be identified, and, most importantly, how the lighting will change needs to be understood. For example, if there are windows in the space, changing sunlight during the day can create transient bright spots in the camera field of view. Depending on the VID system (flame or smoke and manufacturer), the occurrence of a bright spot may lead to false alarms. However, there are numerous techniques to address this issue and avoid false alarms, such as judiciously specifying detection zones, masking out problem areas in the video, and verifying transient conditions over the duration of a preset time delay before alarm.

The acceptable performance of a smoke VID system in the global fire protection system plan needs to be defined relative to its capability in low light or no light conditions. Smoke VID systems do require a minimum level of light to detect smoke. The specific effect of low light on smoke detection performance depends on the light level, the camera and the particular VID system. Some systems can utilize IR emitters to compensate for low light conditions. As low light is a significant issue for smoke detection, a system design should specify the expected light levels, the anticipated changes in light (i.e., periods of time when it may be off or reduced), and how the VID system will operate during the different light levels.

The design of a smoke VID system (as well as any smoke detection technology) must take into account ventilation, which can have a substantial impact since many spaces have large air flow rates. A primary concern of high air flows is dilution of smoke and distortion of the smoke plume, which will make detection more challenging. For spot and aspiration smoke detection systems, designs for spaces with high air flow rates typically consist of closer spacing of detectors/sample points and/or higher alarm sensitivity settings. The relationship between dilution, smoke concentration and spot/air sampling detection is more clearly understood than is the relationship between dilution and smoke concentration relative to VID detection capability. This is partly true since VID algorithms use other parameters than smoke concentration to determine alarm. For instance, parameters may include smoke movement, edge effects, and changes in brightness and contrast. The effect of ventilation on these parameters is much more uncertain and harder to quantify than estimating smoke concentration due to dilution. It is noted for spot detectors that even with a general understanding of the effects of ventilation on smoke dilution, the full impact of ventilation on detection performance is typically unknown and not analyzed for most applications.

Contamination of a VID camera lens can also be a critical issue that degrades detection performance. Fouling can occur because of dirt, oils, and condensation, particularly in industrial applications. Besides the standard temperature and humidity operating requirements, performance requirements for operation in environments that can cause fouling should be established. These requirements can be addressed in generally three primary ways: (1) establishing increased periodic maintenance procedures, (2) preventing fouling, such as the use of a window purge system, and (3) self-diagnostic functions in the VID system that alert users to poor image quality and the need for cleaning. Most VID systems are being designed with self-diagnostic functions to assess video image. For flame detectors, NFPA 72 (5.8.3.2.6) addresses the requirement to provide a means for sustaining window clarity. Similar requirements for smoke detectors do not exist.

4.3 System Variables

System variables can include the type of camera, camera settings, field of view, sensitivity settings, delays, and image settings (i.e., zones, masking, etc.). Many of the system variables, such as hardware and alarm algorithm settings, are captured by approval testing of the VID system. During the approval or listing of a system, the specific equipment and setup parameters are specified along with the detection performance to well characterized fires. The appropriateness of the tested fire detection performance must be compared to the specific application performance objectives. If the conditions tested are not suitably similar to the specific application (e.g., fire type, background conditions, etc.) than additional testing or on-site validation may be needed to satisfy the system design objectives.

Integration of VID systems with other fire alarm, suppression, building or security systems is generally done by standard outputs, such as relay contacts and 4–20 mA signals. These interfaces are also evaluated by listing and approval organizations.

The type of system architecture needs to be considered when defining the performance requirements. If the system is a collection of cameras with individual video feeds to a central processing unit, there will typically be more performance issues to address. Unless, the camera and wiring hardware happens to be the same as specified in the approval of the VID system, the response to fires and potential false/nuisance sources may be different than tested. Self-contained VID system cameras have the advantage of maintaining tighter controls on performance. The type of architecture can also impact how the VID system interfaces with other systems and the extent to which it can be accessed and controlled via remote sources. Performance objectives for interface and control should be specified for a system. For example, some systems can be accessed and controlled via internet at remote sites. The ability to view the video with alarm conditions at off-site locations may be critical for a specific application.

The ability of a system to ignore potential false/nuisance sources can be addressed in multiple ways. Ideally, the VID alarm algorithm is designed to be highly discriminating. The degree of effective discrimination can vary substantially between VID systems. Several VID products have specific system variables that can be used to address false/nuisance sources. These include being able to identify only specific zones within the image for detection. For example, zones could be established in the video image above head height so that all motion of people is below the detection zone. Alternatively, some systems can mask (blackout) areas of the video image within which the alarm algorithms will not be evaluated. For example, an exhaust stack could be masked out to avoid potential nuisance smoke alarms from the exhaust. Another means consists of adjusting the persistence time of the event before an alarm signal is issued (i.e., a preset time delay to avoid brief transient events). Some systems have the means to adjust alarm sensitivity during different times of day to avoid potential false/nuisance events that are more likely to occur during specific hours of operation. The design performance objectives will impact how these means to avoid false/nuisance sources are implemented. Depending on the fire scenario, field of view and required alarm time needed for fire detection, not all of the mechanisms for false alarm rejection may be appropriate. If fast detection is the primary goal, then reduced sensitivity and time delays would not be acceptable methods for minimizing false/nuisance sources.

The performance requirements for VID systems to supervise hardware, functions and field of view must be established for the application. Supervision of hardware depends on the system architecture. Self-contained VID cameras should electrically supervise limited life component failures and loss of power as is evaluated in UL 268 for smoke detectors. VID systems with standard cameras and a central processing unit can supervise power to all elements but cannot directly supervise component failures in cameras. However, they can indirectly supervise the camera component failures if it affects the camera image.

The misalignment of a camera from its intended field of view can be a critical performance issue, particularly if the intended protected area is no longer in the field of view. Establishing a requirement that a VID system must be able to supervise the camera view from being moved or obscured is a reasonable performance criterion. For reference, there are no current requirements (nor capability) for OFDs to independently supervise their field of view. Similarly, there are no requirements that smoke detectors supervise smoke entry capabilities (e.g., a dust cover being left on the device), which would be analogous to a VID system supervising an obscured camera view. However, some VID systems do have capabilities to both monitor camera field of view as well as obscured camera views from either lens contamination or objects placed within the near field of view of the camera. Although these features are advantageous, whether they are necessary is dependent on the application. Most designs seek to securely mount cameras in locations that are unlikely to be disturbed or obstructed.

Similar to alignment and obscuration issues, VID systems typically implement self-diagnostics to determine video image quality for proper detection performance. Besides environmental contaminants, video images may be degraded due to hardware adjustments that change focus and brightness of the image. System reliability is increased by establishing performance criteria that either prevents hardware changes that affect image quality or by requiring software diagnostics to supervise image quality.

4.4 Comparison to Other Detection Systems

A primary difference between VID and smoke (spot or beam) and optical flame detection is that VID is relatively new and unknown to many. However, a close examination of the performance criteria and even many of the design considerations for the use of projected beam smoke detectors and optical flame detectors (OFD) reveals that there is much in common with the design of a VID system. For most of the spacing and installation requirements, NFPA 72 does not differentiate between OFDs and flame VID systems [2]. It establishes a performance-based design approach for the use of OFDs and flame VID systems (Sect. 5.8 Radiant Energy-Sensing Fire Detectors). First, the performance objective of the system must be stated in the design documentation [as is true for any detection system

(NFPA 72 Sect. 5.6.1.1 for heat and 5.7.1.1 for smoke)]. This requirement is obviously application specific. Selection of the flame detector is based on the particular fuel type, the fire growth rate, the environment, the ambient conditions, and the capabilities of the extinguishing media and equipment (NFPA 72, Sect. 5.8.2.1). Besides a fire hazard analysis, the basis for addressing most of these issues is testing conducted by the manufacturer or approval agency. NFPA 72 also states that the flame detection system should minimize the possibility of spurious nuisance alarms from non-fire sources inherent to the hazard area. There are no prescriptive spacing requirements for flame detectors other than that they shall be used consistent with the listing or approval and shall be positioned to assure that all points to be protected are within the field of view of at least one detector. There are several requirements to avoid obstructions within the field of view and to maintain window clarity in dirty environments. As recognized by the NFPA Technical Committee on Initiating Devices, flame VID systems are very similar to other flame detection systems and shall be designed in a similar manner. As with variations between different OFDs, the variations between different VID technologies require application specific design considerations.

Although there are more prescriptive installation requirements for spot smoke detection systems, there is limited data that verifies the actual performance of systems installed per the prescriptive requirements relative to the performance objective of the system design. For example, installing spot detectors using the recommended 9.1 m (30 ft) spacing does not assure timely detection (or detection at all) for all building configurations, ventilation modes and fire type scenarios. For projected beam smoke detectors, the primary criterion for installation per NFPA 72 is that the system shall be located in accordance with the manufacturer's published instructions. However, there is even less published data than for spot detectors that validates the performance of beam smoke detection systems. Many would be hard-pressed to accurately identify (much less to verify) the size of fire that could be detected by a spot- or beam smoke detector given a specific building structure and alarm sensitivity. As such, there is still a need for research in the area of smoke development and measurement in applications in which spot, beam and VID smoke systems would be employed. Smoke VID systems are more similar to projected beam smoke detection systems, for which there are limited prescriptive requirements and no validated tools for calculating detector performance.

Given the unfamiliarity with VID systems, there seems to be a heightened demand for more rigorous testing and performance criteria than currently exists for detection systems commonly in use today. Although it is appropriate for any new technology to be evaluated and challenged until it develops a pedigree of acceptable performance, the demands placed on the system should not be onerous or beyond what is done for other existing systems. NFPA 72 has already applied performance-based design requirements to flame VID systems by including them as just another flame detector. It is recommended that the NFPA 72 technical committees develop a clear framework for the performance-based design approach for using smoke VID systems. However, it is recognized that the knowledge and tools available for performing such designs, for any smoke detection technology, still need to be further improved [16, 17]. One specific recommendation for smoke or flame VID systems is that general geometrical guidelines could be developed for the planning of camera layouts relative to the lens being used and the stated range of the VID system.

Chapter 5
Conclusions

The industry workshop on VID systems as well as a review of fire incident data and multiple occupancies that are using or considering the use of VID systems has clearly revealed the complexity of fire VID applications. Even for similar occupancies or even different camera locations within the same space, there can be large variations in the environmental factors that impact VID performance. In general, background images (e.g., structures and equipment), colors of walls and objects, lighting, and activities within the field of view of a camera can have effects on the operation of a system. This variation does not inherently imply problems for VID systems, but it does create a problem for selecting specific environmental parameters in developing performance criteria that are to be broadly applicable. Consequently, similar to optical flame detectors, the design of a VID system should follow an application-specific, performance-based approach.

The performance of a smoke or flame VID system must take into account three general items:

1. Fire sources.
2. Environment.
3. System variables.

In general, fire sources are defined by the specific application being addressed and the desired response from detection to manual or automatic actions. The environment, such as background, foreground, lighting, contaminants, etc., are dependent on the building and its operations. System variables are dependent primarily on the specific VID system being used; although some aspects, such as field of view, are also dependent on the environment being protected.

For warehouse/distribution centers and large industrial applications, such as power plants and petroleum processing plants, the main goal of VID systems is to provide property protection and mission continuity. In general the fire detection systems are desired to provide early warning necessary for quick, manual intervention. The primary concerns for these occupancies are Class B and Class A

D. T. Gottuk, *Video Image Detection Systems Installation Performance Criteria*, 39
SpringerBriefs in Fire, DOI: 10.1007/978-1-4614-4202-8_5,
© Fire Protection Research Foundation 2008

fuels; however, electrical failure and electrical distribution fires constitute a significant number of incidents in both of these general applications and are pertinent to early warning detection, particularly for smoke production. Especially for the industrial applications, combustible and flammable liquids and gases are a concern. For both types of occupancies, the goal of early detection leads to the performance objective of detecting fires on the order of 25–100 kW.

The same argument for fire size can be made for atriums, for which life safety is the priority goal. Since atriums routinely include the presence of people, providing early detection for evacuation is a main objective. A Class A fire of approximately 100 kW would be a reasonable performance objective given the size and openness of most atriums. The minimum fire size for the activation of smoke management systems in atriums to maintain tenable conditions within the space and adjacent connected spaces will be larger than 100 kW. Therefore, a 100 kW fire provides time for notification of occupants prior to when a smoke management system would operate, after which, by design, occupants should still have time to evacuate.

Because of the ability to have instant video upon alarm, users have indicated a higher level of acceptance of false/nuisance alarms than is typical with typical fire alarm systems. For several applications in distribution centers and electric power plants, users stated that 1–2 alarms per day would be acceptable. Although there can be many factors that can contribute to potential false/nuisance alarms, VID systems have multiple means of dealing with them. Development also continues to address the more common issues with improved discriminating alarm algorithms.

There are multiple environmental and system variables that need to be considered in the design and use of a flame or smoke VID system. These can include obstructions, changing light levels, ventilation, lens contamination, camera settings, and system integration. The operating performance criteria for a system must define the parameters in which the VID system is to operate. Based on these parameters and the main performance objective of the detection system, the means of addressing environmental and system variables can be determined.

Given the unfamiliarity with VID systems, there seems to be a heightened demand for more rigorous testing and performance criteria than currently exists for detection systems commonly in use today. Although it is appropriate for any new technology to be evaluated and challenged until it develops a pedigree of acceptable performance, the demands placed on the system should not be onerous or beyond what is done for other existing systems. NFPA 72 has already applied performance-based design requirements to flame VID systems by including them as just another flame detector. It is recommended that the NFPA 72 technical committees develop a clear framework for the performance-based design approach for using smoke VID systems. However, it is recognized that the knowledge and tools available for performing such designs, for any smoke detection technology, still need to be further improved [16, 17]. For example, the application of beam smoke detectors is done with limited ability to verify the performance objective of the system. One specific recommendation for smoke or flame VID systems is that general geometrical guidelines could be developed for the planning of camera layouts relative to the lens being used and the stated range of the VID system.

Chapter 6
Future Work

Based on this study, the following recommendations for future work have been identified.

1. Given the range of environmental and system variables that can impact a design, it is recommended that a systematic test program be conducted to quantify the effect on VID system performance relative to key variables. With the relatively recent development and growing use of flame and VID systems, there is a general need by end-users and fire protection system designers to have documented, quantitative performance capabilities. These capabilities should include both fire detection and false/nuisance alarm discrimination. One area of concern is detection relative to obstructions between the flame/smoke and the camera. There is a lack of data and testing of fire detection capability (particularly for flame detection) for fires that are partially obstructed in the field of view or flames impinging on objects. In these scenarios, flames will be distorted or discontinuous in the video image compared to fires burning in the open. It is also recommended that systems be tested with cameras viewing the fire from ignition, instead of shuttering the fire exposure until it is developed. It is more realistic to have detection systems view the fire development.

2. It is recommended that performance limitations be developed for all smoke detection systems (VID), spot, beam, and aspiration). Particularly for smoke detection, some of the same variables and fire scenarios that would be evaluated for VID systems would also be applicable for spot and beam smoke detection systems. Despite their longer history of use, there is still more work needed to quantify the effects of key variables on spot and beam smoke detection performance. In general, without advanced modeling, the limitations of performance for spot and beam smoke detection systems is not apparent for the range of applications that they are being used in. Consequently, there is technically just as much of a need for experimentally evaluating spot and beam detection systems as there is for VID systems.

D. T. Gottuk, *Video Image Detection Systems Installation Performance Criteria*, 41
SpringerBriefs in Fire, DOI: 10.1007/978-1-4614-4202-8_6,
© Fire Protection Research Foundation 2008

3. It is recommended that an independent survey of VID performance of installed systems be conducted. This study would allow the development of a database that can be used to establish performance reliability. The results of which would help develop a VID technology pedigree, provide data for assessing performance-based designs, and provide data that can help manufacturers advance the technology by addressing identified issues.

4. It is recommended that additional code language be developed for NFPA 72 that provides for considerations of VID system related performance criteria as discussed in this report. Also a hazard assessment and design methodology should be developed, particularly for smoke VID systems. In general, the methodology in NFPA 72 for the design of radiant energy-sensing fire detection systems is sufficiently broad to cover flame VID systems; however there is some language, particularly in the annex material, that is not necessarily appropriate for VID technology. Although it is recommended to develop a performance-based design methodology for smoke VID systems, the knowledge and tools available for performing such designs, as well as for any smoke detection system, still need to be improved, developed and validated. One example is validating smoke movement modeling relative to ventilation and stratification effects.

Appendix A
NFPA 72-2007 Code Requirements

NFPA 72-2007 Requirements Pertaining to Fire VID Systems

5.7.6 Video Image Smoke Detection

5.7.6.1 Video image smoke detection (VISD) systems and all of the components thereof, including hardware and software, shall be listed for the purpose of smoke detection.

5.7.6.2 VISD systems shall comply with all of the applicable requirements of Chaps. 1, 4–6, and 10 of this code.

> 5.7.6.2.1 Systems shall be designed in accordance with the performance-based design requirements of Sect. 3.3.
>
> 5.7.6.2.2 The location and spacing of video image smoke detectors shall comply with the requirements of 5.10.5.

5.7.6.3* Video signals generated by cameras that are components of VISD systems shall be permitted to be transmitted to other systems for other uses only through output connections provided specifically for that purpose by the video system manufacturer.

5.7.6.4* All component controls and software shall be protected from unauthorized changes. All changes to the software or component settings shall be tested in accordance with Chap. 10.

5.5.8 Video Image Flame Detection

5.8.5.1 Video image flame detection (VIFD) systems and all of the components thereof, including hardware and software, shall be listed for the purpose of flame detection.

5.8.5.2 VIFD systems shall comply with all of the applicable requirements of Chaps. 1, 4–6, and 10 of this code.

D. T. Gottuk, *Video Image Detection Systems Installation Performance Criteria*, SpringerBriefs in Fire, DOI: 10.1007/978-1-4614-4202-8,

5.8.5.3* Video signals generated by cameras that are components of VIFD systems shall be permitted to be transmitted to other systems for other uses only through output connections provided specifically for that purpose by the video system manufacturer.

5.8.5.4* All component controls and software shall be protected from unauthorized changes. All changes to the software or component settings shall be tested in accordance with Chap. 10.

5.10.5 Location and spacing of detectors shall comply with 5.10.5.1 through 5.10.5.3.

    5.10.5.1 The location and spacing of detectors shall be based on the principle of operation and an engineering survey of the conditions anticipated in service. The manufacturer's published instructions shall be consulted for recommended detector uses and locations.

    5.10.5.2 Detectors shall not be spaced beyond their listed or approved maximums. Closer spacing shall be used where the structural or other characteristics of the protected hazard warrant.

    5.10.5.3 The location and sensitivity of the detectors shall be the result of an engineering evaluation that includes the following:

(1) Structural features, size, and shape of the rooms and bays
(2) Occupancy and uses of the area
(3) Ceiling height
(4) Ceiling shape, surface, and obstructions
(5) Ventilation
(6) Ambient environment
(7) Burning characteristics of the combustible materials present
(8) Configuration of the contents in the area to be protected

10.4.3 Video Image Smoke and Flame Detectors. Video image smoke and flame detectors shall be inspected, tested, and maintained in accordance with the manufacturer's published instructions.

10.4.4* Testing Frequency. Testing shall be performed in accordance with the schedules in Table 10.4.4, except as modified in other paragraphs of 10.4.4, or more often if required by the authority having jurisdiction.

A.3.3.181.5 VISD VISD is a software-based method of smoke detection that has become practical with the advent of digital video systems. Listing agencies have begun testing VISD components for several manufacturers. VISD systems can analyze images for changes in features such as brightness, contrast, edge content, loss of detail, and motion. The detection equipment can consist of cameras producing digital or analog (converted to digital) video signals and processing unit(s) that maintain the software and interfaces to the fire alarm control unit.

A.3.3.209 VIFD VIFD is a software-based method of flame detection that can be implemented by a range of video image analysis techniques. VIFD systems can analyze images for changes in features such as brightness, contrast, edge content, loss of detail, and motion. The detection equipment can consist of cameras producing digital or analog (converted to digital) video signals and processing unit(s) that maintain the software and interfaces to the fire alarm control unit.

Appendix B
VID System Workshop

VID Workshop Attendees March 10, 2008

WayneAho	xtralis Inc	waho@xtralis.com
Oded Aron	Port Authority of New York & New Jersey	oaron@panynj.gov
Vito Bellantuono	Underwriters Laboratories Inc.	vito.beJlantuono@us.ul.com
Mark Boone	Dominion Resources Services, Inc	marks.boone@dom.com
David Bradley	xtralis Inc	dbradley@xtralis.com
Thomas Brown	Rolf Jensen & Associates, Inc.	tbrown@rjagroup.com
Dave Christian	Gentex Corporation	davec@gentex.com
Shane Courbier	Gentex Corporation	shane.courbier@gentex.com
John Dolan	AD Group Inc	john.dolan@ad-groupinc.com
Curtis Donahou	Walt Disney World Resort	curtis.donahou@disney.com
Kenneth Dungan	PLC Foundation	kwdungan@risktek.com
Robert Elliott	FM Approvals	robert.elilott@fmapprovals.com
Thomas Fabian	Undervvriters Laboratories Inc	Thomas Fabian@us ul com
Alan Finn	United Technologies Research Center	FlnnAM@utrc.utc.com
Bruce Fraser	Fraser Fire Protection Services	bruce@fraser-fpscom
Stuart Fuller	AD Group Inc	stuart.fuller@dtec-analytics.com
Kenneth Gentile	The RJA Group, Inc	kgentile@rjagroup.com
James Glockling	Fire Protection Association	jglockling@thefpa.co.uk
Daniel Gottuk	Hughes Associates, Inc.	dgottuk@haifire.com
Daniel Grosch	Underwriters Laboratories Inc.	daniel.m.grosch@us.ul.com
Stewart Hall	Tyco Safety Products	stewarthall@tycoint.com
David Icove	The University of Tennessee	icove@utk.edu
Jon Kapis	Rolf Jensen & Associates, Inc.	jkapis@rjagroup.com
Ron Knox	xtralis Ai Ltd.	rknox@xtralis.com
Scott Lang	System Sensor	scott.lang@systemsensor.com
Rick Lewis	Rolf Jensen & Associates	rlewis@rjagroup.com

(continued)

D. T. Gottuk, *Video Image Detection Systems Installation Performance Criteria*, SpringerBriefs in Fire, DOI: 10.1007/978-1-4614-4202-8, © Fire Protection Research Foundation 2008

(continued)

Adrian Lloyd	Micropack Detection (Americas) Inc.	AdrianLloyd@micropackamericas.com
David Lloyd	Fire Sentry Corporation	dlloyd@firesentry.com
Andy Lynch	axonXLLC	ancfy@axonx.com
Greg Masterson	Liberty Mutual Property	gregory.masterson@libertymutual.com
Max McLeod	Siemens Building Technologies, Inc.	max.mcleod@siemens.com
Jack McNamara	Bosch Security systems	jack.mcnamara@us.bosch.com
Thomas McNelis	Intertek ETL SEMKO	thomas.mcnelis@intertek.com
Noura Milardo	FM Global	noura.milardo@fmglobal.com
Wayne Moore	Hughes Associates, Inc	wmoore@haifire com
Mac Mottley	axonXLLC	mmottley@axonx.com
Curtis Nance	Space Age Electronics	curtis.nance@1sae com
Louis Nash	US Coast Guard	louis.nash@uscg.mil
Dave Newhouse	Gentex Corporation	daven@gentex.com
Ronald Ouimette	Siemens Building Technologies, Inc	Ronald.Ouimette@siemens com
Isaac Papier	Honeywell Life Safety	isaac.papier@honeywell.com
John Parssinen	Underwriters Laboratories Inc	john.l.parssinen@us.ul. com
Paul Patty	Underwriters Laboratories Inc	paul.e.patty@us.ul com
Raymond Quenneville	FlreFlex Systems Inc	rquennevllle@fireflex.com
Rodger Reiswig	SimplexGrinnel1	rreiswig@tycoint.com
Jeffrey Roberts	XL Global Asset Protection Services	Jeffrey.Roberts@xlgroup.com
Sue Sadler	xtralis Inc.	ssadler@xtralis.com
Robert Schifiliti	R.P Schifiliti Associates, Inc	rps@rpsa-fire com
Ian Thomas	Victoria University	Ian.Thomas@vu.edu.au
Daniel Wepfer	Durag GmbH	daniel.wepfer@durag.de

Questions	Warehouse/Distribution centers
1. Life safety, property protection, mission continuity?	Property protection and mission continuity
2. How will detection and video be utilized in the fire protection system? • Manual response? • Activation of other systems? • Need for pre-alarm signals? • Human interface requirements? • Need for video verification? How is it to be used? • Alarm nuisance	Detection depends on what you're protecting; automatic suppression combined w/activation of system. All the utilizations can be justified depending on the application Not just detection for protection activation, but also as video verification for zone evacuation purposes Especially for MNEVAC mass notification False alarm key to high end inventory warehouse
3. Need for smoke detection?	Not a code requirement. For high-end storage, smoke detection matters, otherwise, unless detection ties into extinguishing, smoke detection not necessary
4. Need for flame detection?	Yes—flames mean ignition/fire. Limitation due to view vantage. Smoke detection VID captures this, but also view obstruction for product-type commodities. Base VID on burning characteristics of commodity.
5. Need for security surveillance?	
6. Expected types of fires (gas, Class A, Class B, pool, spray…)	Warehouse aisles vary; commodities vary; heights of storage; cold storage warehouses; automated storage—all these variables change need of detection and type of fire to expect.
7. Desired fire size for detection threshold?	20–100 Kw for smoke, higher for flame–300 Kw? Researchers should figure this out. Output rate should be heat release rate (HRR).
8. Potential nuisance sources?	Change of ambient light (doors), dust, steam, obstructions, space heaters, condensation, field of view, sparks from controlled fire, exhaust from fork lift. Severity of nuisance alarm and # to contend with.
9. System architecture (e.g., self-contained devices or distributed cameras with central processing unit)?	Technology allows for self contained systems, so no need to have separate components. For now, approvals in place ask for complete supervision and reliability.
10. System integration with other building, security and fire alarm systems?	Combination FACP, VID, and Security—as long as all components are reliable and listed/approved.
11. Type of outputs desired (4-20 mA, dry contacts, IP communications…)?	Desire for pan view, so IP communication (dedicated networks) is mostly desired for cost effectiveness; dry contact works here
12. What information needs to be communicated?	Location
13. Video camera attributes needed (pan zoom tilt, black and white, color…)?	Black & white processing with colored cameras

(continued)

(continued)

Questions	Warehouse/Distribution centers
14. Desired camera locations?	Need to avoid obstructions. Aerosol storage, pallet storage—over racks Battery storage location
15. Camera accessibility issues?	Size of warehouse and type of fire determine where to put cameras and how to reach them
16. Camera contamination issues?	Dust accumulations—keep VID systems in locations that don't cause dust
17. Listing and approval requirements for this occupancy? Source of requirement?	Depends on agency doing the listing; as long as its performance is reliable and special application
18. Sizes of protected spaces (area and height)?	Up to 40' ceiling, greater than 1 million sq ft. For flame detection, each camera could protect ~ 140 ft with a 90° angle camera
19. General layout (large open space, highly congested, aisles, open space above equipment)?	Large open space, not congested; above aisles is preferable; supported by mapping tools
20. Desired field-of-view (depth and angle)?	Flexibility is important since warehouse can be quite large/tall
21. Operations within the field-of-view (mechanical, vehicles, people...)?	Lots of movement—need to accommodate for that
22. Challenges of occupancy configuration/ obstructions to line of sight detection?	Narrow aisles; obstructed aisles
23. Expectation that field-of-view image will change (how often, in what ways, transient motions (seconds or minutes) or semi-permanent (hours to days))?	If cameras are high enough, this is not an issue
24. Visual environment • Background color? • Background pattern (objects ...)?	Changing background, color, objects –cameras/ processing needs to adjust
25. Lighting source?	This is necessary—no viewing ability w/out lighting for smoke; flame is ok
26. Lighting schedule?	N/A—All night, still need light. No issue for flame VID
27. Light levels w.r.t. areas of protection?	Same as above. Systems have capability to report trouble such as loss of light
28. Environmental conditions: • Temperature? • Moisture? • Shock? • Vibration? • Electrical considerations (EMI requirements)?	End users need to be sure that VID systems have been tested and approved for use under all these conditions. For warehouses w/cold storage, moisture and humidity issues. Pharmaceutical facility can benefit here. Mostly data storage—telecom facilities
29.	
30.	
31.	
32.	
33.	
34.	
35.	

Questions	Large industrial applications
1. Life safety, property protection, mission continuity?	1) Mission continuity and 2) Property protection 3) Life safety a future need
2. How will detection and video be utilized in the fire protection system? • Manual response? • Activation of other systems? • Need for pre-alarm signals? • Human interface requirements? • Need for video verification? How is it to be used?	Early warning necessary for quick, manual intervention
3. Need for smoke detection?	Combination smoke/flame detection needed.
4. Need for flame detection?	Combination smoke/flame detection needed.
5. Need for security surveillance?	Combined systems are a positive—must be reliable.
6. Expected types of fires (gas, Class A, Class B, pool, spray...)	– Smoke, flame, depends on the application – Class A/B typical; Class C/D possible – Exterior applications of equal importance
7. Desired fire size for detection threshold?	7 x 7 pixels
8. Potential nuisance sources?	Dirt, oils, humidity, light, etc.
9. System architecture (e.g., self-contained devices or distributed cameras with central processing unit)?	Integrated systems desired; individual systems typically operate as managed Secondary power necessary for life safety occupations —camera signal integrity—UL listings needed
10. System integration with other building, security and fire alarm systems?	– Integrated systems desired; individual systems are managed by the integrating software platform – System capability
11. Type of outputs desired (4–20 mA, dry contacts, IP communications...)?	– I/P based, 4–20 mA – Supplemental
12. What information needs to be communicated?	– Size, location of fire, response needed, etc. – Post
13. Video camera attributes needed (pan zoom tilt, black and white, color...)?	– Future needs—primarily in security currently – Horizontal view
14. Desired camera locations?	– Horizontal view
15. Camera accessibility issues?	– Accessible for maintenance & testing
16. Camera contamination issues?	– Housings should protect in maintained environment
17. Listing and approval requirements for this occupancy? Source of requirement?	– UL Listing needed; NFPA 72 – NEC
18. Sizes of protected spaces (area and height)?	– Based on field of view
19. General layout (large open space, highly congested, aisles, open space above equipment)?	– Ventilation, high air flow; all listed spaces are possible layouts

(continued)

(continued)

Questions	Large industrial applications
20. Desired field-of-view (depth and angle)?	– Horizontal view, possible multi-camera views – Flexible installation options for location
21. Operations within the field-of-view (mechanical, vehicles, people...)?	– All evident
22. Challenges of occupancy configuration/ obstructions to line of sight detection?	– Application dependent
23. Expectation that field-of-view image will change (how often, in what ways, transient motions (seconds or minutes) or semi-permanent (hours to days))?	
24. Visual environment Background color? Background pattern (objects ...)?	
25. Lighting source?	
26. Lighting schedule?	
27. Light levels w.r.t. areas of protection?	
28. Environmental conditions: • Temperature? • Moisture? • Shock? • Vibration? • Electrical considerations (EMI requirements)?	
29.	
30.	
31.	
32.	
33.	
34.	
35.	

Questions	Atriums—suggest the presence of people/life safety
1. Life safety, property protection, mission continuity?—Continue business function as ususal	Priority: (1) Life safety—provide plausible egress! (2) Security (3) Mission continuity/people protection
2. How will detection and video be utilized in the fire protection system? • Need for pre-alarm signals? ——— • Human interface requirements? —— • Need for video verification? How is it to be used? ——— • Manual response?—Fire Brigade • Activation of other systems? ———	Only a fire in the actual atrium will VID be helpful (as opposed to connected areas) VID must be listed as a smoke detector in order to use – Yes, 'investigation period' needed—(verify/confirm alarm validity) "Positive Alarm Sequence" – Yes, allow for false alarms and judgment calls based on specific scenario/conditions Via 'select zones'—allow for environment changes (population moving/changing constantly in atria) – Smoke Control/HVAC—FACP reporting is a code requirement—suppression system communication
3. Need for smoke detection?	Yes—VID must be listed as a smoke detector to use
4. Need for flame detection?	Depends
5. Need for security surveillance?	Insure goals/objectives are well established in advance. Permissible—fire is priority; a supplementary system not applicable
6. Expected types of fires (gas, Class A, Class B, pool, spray...)	Class A
7. Desired fire size for detection threshold?	TBD—engineering analyses needed for subject space Fire must be discernable to viewer—distance placement of camera is a factor—to include space immediately under camera as well as any other 'blind spots'
8. Potential nuisance sources?	Visual verification will keep these to a minimum
9. System architecture (e.g., self-contained devices or distributed cameras with central processing unit)?	
10. System integration with other building, security and fire alarm systems?	
11. Type of outputs desired (4–20 mA, dry contacts, IP communications...)?	
12. What information needs to be communicated?	
13. Video camera attributes needed (pan zoom tilt, black and white, color...)?	No PTZ (pan-tilt-zoom)

(continued)

(continued)

Questions	Atriums—suggest the presence of people/life safety

14. Desired camera locations?
15. Camera accessibility issues?
16. Camera contamination issues?
17. Listing and approval requirements for this occupancy? Source of requirement?
18. Sizes of protected spaces (area and height)?
19. General layout (large open space, highly congested, aisles, open space above equipment)?
20. Desired field-of-view (depth and angle)?
21. Operations within the field-of-view (mechanical, vehicles, people...)?
22. Challenges of occupancy configuration/ obstructions to line of sight detection?
23. Expectation that field-of-view image will change (how often, in what ways, transient motions (seconds or minutes) or semi-permanent (hours to days))?
24. Visual environment
• Background color?
• Background pattern (objects ...)?
25. Lighting source?
26. Lighting schedule?
27. Light levels w.r.t. areas of protection?
28. Environmental conditions:
• Temperature?
• Moisture?
• Shock?
• Vibration?
• Electrical considerations (EMI requirements)?
29.
30.
31.
32.
33.
34.
35.

Consensus—in many instances and for a variety of reasons, atria may not be ideal environments for VID for primary detection, though they perhaps lend value as a supplementary system. Constant changing of environment—human element—connected to other and common areas in which we could only speculate and theorize as to actual scenario and code requirement imposed on those areas, i.e., detection versus suppression.

References

1. Gottuk DT, Lynch JA, Rose-Pehrsson SL, Owrutsky JC, Williams FW (2004) Video image fire detection for shipboard use. AUBE '04—proceedings of the 13th international conference on automatic fire detection. Duisburg, Germany, 14–16 Sept 2004
2. NFPA 72 (2007) National fire alarm code. National Fire Protection Association. Quincy, Massachusetts
3. Zakrzewski RR, Sadok M, Zeliff B (2004) Video-based Cargo fire verification system for commercial aircraft. AUBE '04—proceedings of the 13th international conference on automatic fire detection. Duisburg, Germany, 14–16 Sept 2004
4. Leisten V, Bebermeier I, Oldorf C (2004) Camera-based fire verification system. AUBE '04—proceedings of the 13th international conference on automatic fire detection. Duisburg, Germany,14–16 Sept 2004
5. Krull W, Willms I (2004) Test methods for a video-based Cargo fire verification system for commercial aircraft. AUBE '04—proceedings of the 13th international conference on automatic fire detection. Duisburg, Germany, 14–16 Sept 2004
6. ANSI/FM 3260 Radiant energy-sensing fire detectors for automatic fire alarm signaling. Approval standard, factory mutual, class no. 3260, Feb
7. UL 268 (2001) Smoke detectors for fire alarm signaling systems. 5th edn. Underwriters Laboratories Inc., Northbrook, 8 Sept 2006
8. Babrauskas V (2002) Heat release rates. Section 3/Chapter 1, The SFPE handbook of fire protection engineering, 3rd edn. In: DiNenno PJ (ed)
9. Ahrens M (2006) Storage properties excluding dwelling garages. National Fire Protection Association
10. Rohr KD (2002) Steam, heat energy and electric generating plant fires and explosions. National Fire Protection Association
11. Twomey ER (2006) Power plant structure fires by alarm year: 1980–2002 and power plant fire incidents with a reported estimated loss of five million dollars or More: 1990–2004. National Fire Protection Association, May 2006
12. Ahrens M (2008) Structure fires in electric generating plants. National Fire Protection Association
13. Ahrens M (2002) Fires in petroleum refineries and natural gas plants 1994–1998 annual averages and trends since 1980. National Fire Protection Association
14. Gottuk DT (2008) International road tunnel fire detection research project—tasks 5 and 6: monitoring and fire demonstrations in the Lincoln tunnel. The Fire Protection Research Foundation, Quincy

D. T. Gottuk, *Video Image Detection Systems Installation Performance Criteria*,
SpringerBriefs in Fire, DOI: 10.1007/978-1-4614-4202-8,
© Fire Protection Research Foundation 2008

15. Ferreira MJ (2008) Fire dynamics simulator, ensure your software provides the safest atrium design for real world enforcement. NFPA Journal, National Fire Protection Association, Quincy, 102(1), January/February 2008

16. Mealy C, Riahi S, Floyd J, Gottuk D (2008) Smoke detector spacing requirements complex beamed and sloped ceilings. Volume 1: experimental validation of smoke detector spacing requirements. Fire Protection Research Foundation, Quincy, April 2008

17. Floyd J, Riahi S, Mealy C, Gottuk D (2008) Smoke detector spacing requirements complex beamed and sloped ceilings. Volume 2: modeling of and requirements for parallel beamed, flat ceiling corridors and beamed, sloped ceilings. Fire Protection Research Foundation, Quincy, April 2008